STRANGE
SEA
CREATURES

Note: On page 48 there is a list of some of
the words in this book and how to say them.

STRANGE SEA CREATURES

By Gina Ingoglia
Illustrated by Turi MacCombie

A GOLDEN BOOK • NEW YORK

Western Publishing Company, Inc., Racine, Wisconsin 53404

Water covers most of the earth. Most of the water is called the sea. The sea is huge. The sea is salty. It may all look the same. But it is not!

Some seawater is warm as bathwater. Some seawater is cold as ice. Parts of the sea are very deep. Parts are not deep at all. But animals live in all parts of the sea. And some of these creatures are pretty strange!

Once people thought there were sea monsters.
They said they saw monsters with long, waving
tails. Now we know sea monsters do not exist.
What sea creatures did people really see?

Maybe they saw a **giant squid**. Giant squids really do exist. These creatures live in the sea. They live in very deep water. Squids have long, wavy arms with suckers on them. Maybe people thought these arms were sea-monster tails.

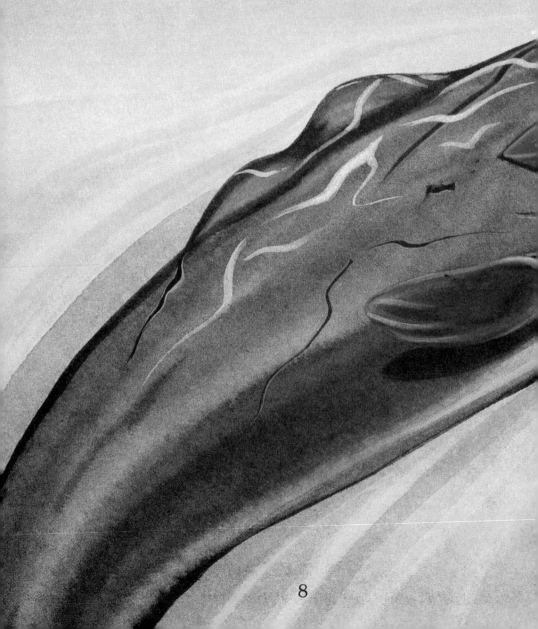

Sometimes giant squids fight with giant sperm whales. What a fight! The squid grabs the whale with its strong suckers. Sometimes the sperm whale gets big marks on its back from the fight. The marks stay on the whale all its life.

Long ago, sailors thought they saw a strange
sea creature. They called it a mermaid. The top
half of a mermaid is a woman. The bottom half is
a fish. Now we know there are no mermaids.
What did the sailors really see?

Some people think the sailors saw a **sea cow.**
Sea cows really live in the sea. These huge
animals eat water plants—sixty pounds of plants
every day! Is this what the sailors saw? Do you
think a sea cow looks like a mermaid?

All sea creatures have bodies that help them live in water. Fish don't breathe air the way land animals do. They stay underwater all the time. Their bodies are shaped for swimming. They have fins and tails to help them swim very fast.

Sometimes the **shark** swims with its large fin sticking out of the water. Most sharks have lots of very sharp teeth. They eat all kinds of sea animals. All sea creatures watch out when a shark is around!

These fish can fly! They have large fins that open into wings. **Flying fish** glide through the air on their open fins. They can get away from other sea creatures that want to eat them.

Flying fish can stay in the air for thirteen seconds. Some flying fish have two fins. Others have four fins. Some flying fish have fins with beautiful colors.

The **manta ray** has fins that look like huge wings. It looks like it has horns coming out of its head. This sea creature can grow twenty feet wide. That's as wide as a small house!

Manta rays can jump out of the water, too. Sometimes they flip over in the air. When they land, they hit the water with a huge, noisy splash.

Sea horses are strange little fish with tiny fins. They bob along as if they are standing on their tails. Their tails are very curly. Sea horses hold on to seaweed with their tails.

Scallops are sea creatures that live between two shells. Peeking out between the edges of the shells are rows of tiny blue eyes. Scallops move in little hops by opening and closing the shells. When the scallop opens its shells, water goes in the front. When it snaps its shells shut, water shoots out the back. As the water shoots out, the scallop is pushed ahead.

These sea creatures stay in one place. They are **coral.** A coral looks more like a plant than an animal. Many coral animals live together on a large coral reef. The reef looks like a very big rock. But it is really all made of coral animals! Some are living. Some are dead.

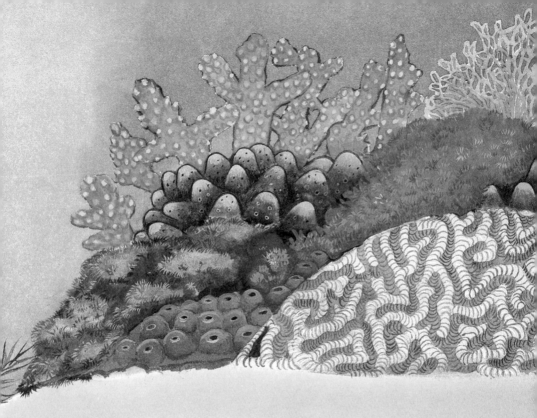

A coral has a tiny body that looks like jelly. It builds a hard tube around itself. One end is stuck to the reef. The other end of the tube is open. The coral has a mouth that sticks out of the open end. Waving feelers bring tiny sea animals called plankton into the mouth. Some large kinds of coral even eat small fish. When the coral dies, the hard tube that covered its body stays stuck on the reef.

Some sea creatures are shaped like stars. They are called sea stars or **starfish.** Some starfish are bigger than dinner plates. Others are smaller than your fingernail.

A starfish has five arms. Under each arm are rows of tiny tube feet. Each tube has a little sucker on its tip. The starfish uses its tube feet to crawl along the sea bottom.

The starfish also holds on to food with its tube feet. Its mouth is under its body.

Most starfish like to eat clams. The starfish puts its arms around the clam and pulls the shells open. Next the starfish does something strange. It sticks its stomach out of its mouth and into the clam! Then the starfish has a clam dinner.

If a starfish loses an arm in a fight, it grows a new one!

This fish fishes for its food. The **angler fish** has a bit of skin hanging over its mouth. It looks just like a wiggly worm. The angler fish lies with its mouth open. A hungry fish sees the "worm" and swims over to eat it. *SNAP!* The angler fish shuts its mouth and *it* gets a meal instead!

Sea anemones look like flowers. But they are animals that eat fish. Their waving arms have poison in them. After a fish is caught, the anemone pulls it into its body and eats it.

But the clown fish is safe. It swims in the anemones and the poison can't harm it.

A **lobster** has a hard shell on the outside of its body. When a lobster get too big for its shell, it sheds it and grows a new one. Some lobsters live in cold water. They have big claws that open and close. They use them to grab animals they want to eat. Lobsters also use their claws to fight.

Some lobsters live in warm water. They don't have claws like cold-water lobsters. Every fall these lobsters take a walk. They walk in long lines on the bottom of the sea. The lobsters may walk more than sixty miles to look for food and mates.

Sea snakes live in warm water. They have a flat tail, like a paddle. They push on the water with it and swim fast. Sea snakes need air to live. With just one gulp of air, they can stay underwater eight hours!

Land snakes shed their skin by rubbing on
trees and rocks. But sea snakes twist up and rub
against their own bodies. Their bite is full of
strong poison.

An eel is a long fish that
looks like a snake.
A **moray eel** lives in
underwater caves and
near coral reefs.
It has sharp teeth
and powerful jaws.

The moray eel looks scary because it is always opening and closing its large mouth. But this is how the eel breathes. As the water goes into its mouth, it passes across slits called gills. The gills take out air that is in the water.

Moray eels may look scary but they bite only to eat or to protect themselves.

Giant sea turtles are small when they hatch from eggs their mothers lay on sandy beaches. After the baby turtles hatch, they swim out to sea. Male turtles never go back to land. Female sea turtles swim to shore only to lay eggs. Many grown giant sea turtles weigh about four hundred pounds. But the biggest ones can weigh over one thousand pounds!

The **grouper** is a fish with a huge mouth. It sucks in gallons of water and any sea creatures that are close by.

The deeper in the sea the grouper lives, the bigger it grows. Some are as big as small cars. They can swallow giant sea turtles!

There are many stories of groupers attacking people. To be safe, divers always watch out for groupers.

Like a sail pushed by the wind, a **man-of-war** gently floats on the sea.

The man-of-war looks like shiny, clear blue glass. It is filled with a gas that makes it float. Long, bumpy strings hang from it. They are full of poison. The man-of-war uses them like a fishing net. When fish bump into the strings, they are trapped!

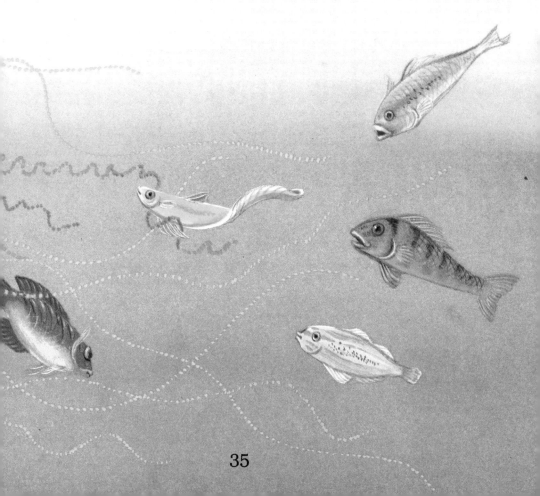

This fish can change shape. When the **porcupine fish** is in danger, it swells up. Sharp spines stick out all over its body. No sea animal wants to eat a fish like that!

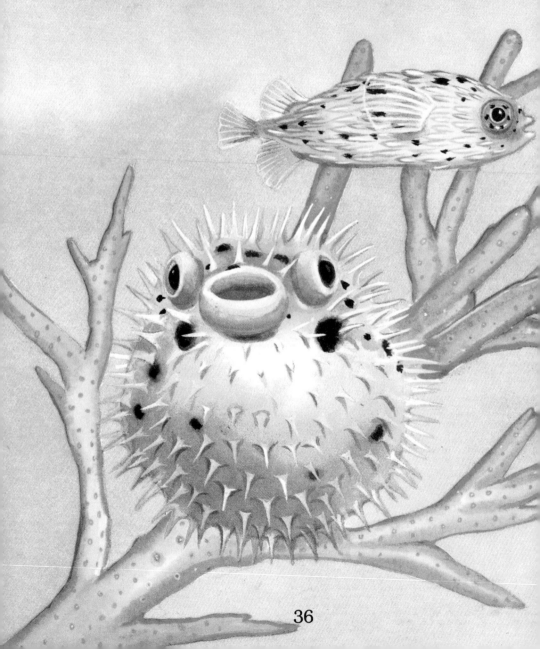

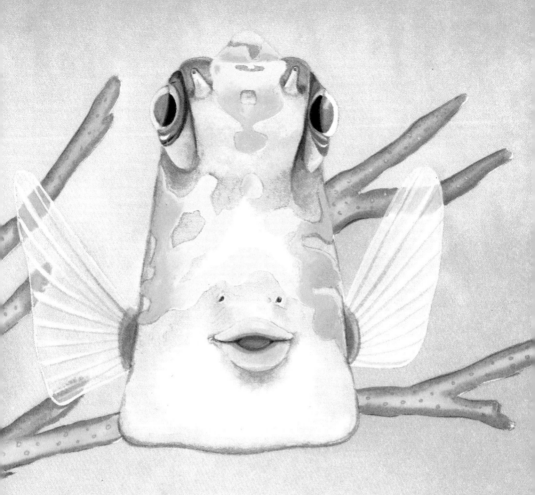

The **cowfish** doesn't have to change shape. It is already hard to eat. Its hard body is shaped something like a box.

The cowfish has a head that looks like the head of a cow—horns and all. But the cowfish has horns with poison in them!

The **stonefish** hides on the bottom of the sea. Its bumpy body looks just like a rock. Bits of coral and tiny water plants stick to it. This fish is also covered with sharp spines. They are full of strong poison.

The little **goby fish** lives in a hole on the sandy sea bottom. It lives with a tiny sea creature called a **shrimp.** The shrimp can't see very well. The goby fish has good eyesight. The fish and shrimp help one another. The shrimp digs the hole. The goby looks out for fish that might eat them.

A **flounder** is as flat as a pancake. It lies on the bottom of the sea. Both eyes are on the top side. A baby flounder looks like most fish, with an eye on each side of its head. But as the fish grows, one eye begins to move. It moves to the other side of its head!

Flounders can change color. When they are lying on sand, they turn a sandy color. They can even look like stones. When a flounder is hiding, it is very hard to find!

In 1938 a fisherman found a strange large
fish in his net. He never saw one like it. It turned
out to be a great find! The fish was a **coelacanth.**
People thought there were no more coelacanths.
They thought coelacanths hadn't lived on earth
for millions of years. But they were wrong.
Coelacanths are still found in the sea today.

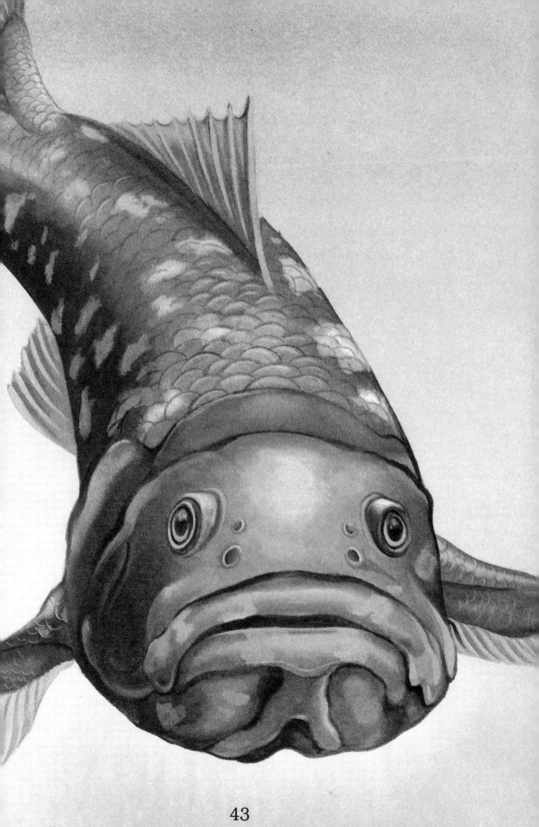

People are learning more and more about sea creatures. Scientists can go deep down in the sea, where it is dark as night. They use strong lights to see. Even sunlight can't get that far down.

But the sea is huge and there are so many places to look! No matter how hard we try, we may never find all the creatures that live in the sea.

The sea creatures in this book live all over the world. But you don't have to go all over the world to see them.

You can see them at a zoo for sea animals called an aquarium. Aquariums are in many cities. You may be lucky. Maybe there is an aquarium not far from where you live!

Here are some of the words from this book and how to say them.

Anemone	(ah NEM mun ee)
Aquarium	(ac QUARE ee um)
Coelacanth	(SEEL lah canth)
Coral	(CORE rul)
Creature	(CREE chur)
Flounder	(FLAOWN der)
Goby	(GO be)
Grouper	(GROO per)
Mermaid	(MUR made)
Poison	(POY zin)
Porcupine	(PORE kew pine)
Scallop	(SCAL up)
Sperm Whale	(spurm wayle)
Spine	(spyne)
Squid	(skwid)